Robert Edmond Grant (1793-1874) and his Museum of Zoology and Comparative Anatomy

by
Dr Sarah E. Parker

Editor
Dr Helen Chatterjee

Associate Editor
Dr Joe Cain

ISBN 0-9552191-0-8
ISBN 978-0-9552191-0-8

Published by
The Grant Museum of Zoology, UCL

Robert Edmond Grant (1793-1874)

The history of Dr. Grant is the history of his published works, his lectures, his connextions with learned Societies, and his scientific explorations... Most retiring in his manners, in his deportment quiet and modest to an unequalled degree, clad always with the most extreme simplicity, - few to whom he might be personally unknown would, from his very unostentatious bearing and appearance, suppose that they beheld a man of his profound knowledge and deserved renown, - until, the hat being removed, the whole contour of his head, his intelligent eye, and the entire character of his countenance, betokened that they saw no ordinary person.

(Anonymous, 1850:695)

Introduction

Robert Edmond Grant was recognised as a man of "profound knowledge and deserved renown" in his day. He was the first professor of Zoology and Comparative Anatomy in Great Britain and, as such, was in a unique position to influence the course of those subjects as academic disciplines in their own right. He was active not only within the burgeoning zoological sciences in London but also played a role in the movement for medical and institutional reform, often at considerable cost to himself, both personally and professionally. He was respected as a Zoologist both at home and in the rest of Europe and was admired by his students. Yet, in spite of his prominent role in the 19th century scientific community, Grant has, until recently, been one of the most under-appreciated figures in the history of science. He is most frequently mentioned as Charles Darwin's first mentor, featuring in little more than a footnote to the history of evolution theory, and usually merits barely a paragraph in Darwin's biography. Consequently, the important part that Grant had to play in the science and teaching of Zoology in 19th century London is not widely understood.

Within the last 20 years, however, our understanding of Grant has increased dramatically. This is thanks to two influences in particular. The first is the detailed examination of Grant and

Darwin's relationship at the University of Edinburgh (Sloan, 1985; Hodge, 1985). In recent years historians of science have attempted to escape the obscuring shade of "Darwin's shadow" (Endersby, 2003). That shadow has threatened to eclipse the influence of other 19th century scientists, placing them only in the context of their impact on Darwin, and not in their own terms alone. The second is the work done by Adrian Desmond (1989) on Grant's work in the context of medical reform in 19th Century London. Desmond presents an extensive examination of Grant's life and work in relation to the social and political climate of the time. He refreshes and brings alive the debates which raged concerning the transformist theories that Grant advocated, debates which extended far beyond the academic arena into the political. Yet, in spite of the detailed treatments of these aspects of Grant's life, there is still much about the man that we have yet to discover.

This booklet is a step towards rectifying the paucity of published information relating to Grant. It presents an accessible introduction to his life and his Museum. The format is largely biographical in order beginning with Grant's early life and interests, moving on to his time at the University of London. The foundation, uses and organisation of his museum, and some background on the more well-known of his students that might have used it will also be discussed. Robert Grant had a staggering depth of knowledge and an influence that is not yet fully recognised. In bringing this remarkable man to the attention of a wider audience it is hoped that this work will help to restore Grant to his rightful place in the history of 19th Century biological sciences.

Grant as a young man

Little is known about Grant's childhood and early life. What there is can mostly be found in a biographical sketch of Grant published in the *Lancet* (Anonymous, 1850). The sketch was part of a series of biographies of prominent scientific and medical figures of the day. Given Grant's close relationship with the *Lancet's* editor – Thomas Wakley (1795-1862) – and the reproduction in that biographical sketch of a piece of personal correspondence sent to Grant by Etienne Geoffroy Saint-Hilaire (1772-1844), it seems reasonable to surmise that the sketch was written with Grant's full co-operation and that many of the facts will have been furnished by Grant himself. As such it provides us with a unique and sympathetic insight into many of the different facets of Grant's life. In addition, there are tantalising tit-bits of information in the various recollections of former students and colleagues of Grant's published during and after his lifetime (Anonymous, 1874-1875; Poore, 1901; Schäfer, 1901).

Robert Edmond Grant c. 1840.
© The Grant Museum of Zoology, UCL.

The seventh son of Alexander Grant, a successful writer to the signet,[1] and Jane Edmond, Robert Edmond Grant was born in Edinburgh on the 11th November 1793. He had eleven brothers and two sisters but was the only one of the twelve sons who settled in Great Britain, his other surviving brothers taking up professions in the Army, Navy or for the East India Company. Grant, however, was destined for the medical profession on account of his more "delicate constitution" (Anonymous, 1850: 689). As a young man, Grant was educated primarily at home by his family's private tutor and later at the High School of Edinburgh. As a youth his interests were initially in Greek and

Geometry and his love of languages continued throughout his life. However, it was when he went to Edinburgh University that his interests began to lean towards the natural world, and the Animal Kingdom in particular, as he attended the Natural History lectures of the geologist Robert Jameson (1774-1854).

As his studies progressed Grant's interests turned away from the classics, which he had originally gone to the University to study, and turned towards medicine. This was a common route for those interested in Natural History at that time and many 19th Century zoologists and botanists were originally trained as medical men – indeed, many of them still practiced medicine in addition to their research. In 1814 Grant graduated as an M.D., writing his dissertation on the circulation of the blood in the foetus (Grant, 1814). Yet, despite his medical qualification, Grant had already resolved to follow a quite different path – he was already bewitched by the study of Comparative Anatomy, a subject that would occupy him for the rest of his life.

After graduating, Grant determined to "[extend] his professional and general knowledge" by studying in the universities, museums and hospitals of continental Europe funded by the inheritance left to him by his father in 1808. Beginning in London, he travelled through France, Italy, Germany,

Switzerland and Austria. He also took a walking tour of Hungary and is said to have crossed the Alps seven times and the Apennines four times on foot before 1820 (Anonymous, 1850). Grant continued to travel throughout his life spending time in Holland and Denmark as well as the rest of Europe. He could speak Dutch, Danish, French, German and Italian and was proficient in Ancient Greek. It was Paris, however, that we could think of as Grant's ideological home.

Grant in Paris

Grant's time in Paris had perhaps the most profound influence on his developing ideas on Comparative Anatomy. French comparative anatomists were some of the most advanced and influential in the world at that time and, while studying there, Grant absorbed many ideas which were considered genuinely revolutionary in his native Great Britain. Grant was a committed transformist having been introduced to such evolutionary ideas while still in Edinburgh by the works of Erasmus Darwin (Grant, 1861). He

Erasmus Darwin. Reproduced with permission. © Wellcome Library, London.

Jean Baptiste Lamarck.
Reproduced with permission.
© Wellcome Library, London.

was also a staunch follower of Jean-Baptiste Lamarck's theories and while in Paris he met and struck up friendships with some of the foremost thinkers of the day. He attended the soirées of Georges Cuvier, later introducing a young Richard Owen to Cuvier's brother Frédéric (also a noted zoologist) while Owen was visiting Paris in 1831 (Desmond, 1989: 57-58). Cuvier offered Grant support and unlimited access to specimens at the Muséum National d'Histoire Naturelle and even forwarded Grant information in order that he might complete a biography of Cuvier (Grant, 1830b).

Though Grant enjoyed the patronage of this giant of French Comparative Anatomy he ultimately rejected Cuvier's theories in favour of a more evolutionary approach. In particular he rejected Cuvier's *embranchements* – the division of the Animal Kingdom into four distinct, unrelated branches: the vertebrata, mollusca, articulata and radiata. He also rejected Cuvier's emphasis on function over form in the study of Comparative Anatomy. He adopted instead the theories of another Parisian comparative

Georges Cuvier. Reproduced with permission. © Wellcome Library, London.

anatomist: Geoffroy Saint-Hilaire. Geoffroy believed in a unity of plan in nature, stretching from the simplest organisms, the monads, to the most complex, human beings. He believed that one could trace that plan by observing the homologies between animals, looking at the forms of the various parts of their bodies rather than their function (see for example, Geoffroy, 1818). That is, one could see that one animal transformed into another because one could see the bones, or organs, in the same places in different animals, even if those bones, or organs, had a different function in each animal. To use one of Geoffroy's examples, the bones of a bony fish's gill cover can be seen in man as the three bones of the inner ear. It was this transformist science that Grant brought with him back to Edinburgh and thence to the University of London.

Etienne Geoffroy Saint-Hilaire. Reproduced with permission. © Wellcome Library, London.

Grant in Edinburgh

Grant returned to Edinburgh in 1820 and quickly devoted himself almost entirely to the study of Comparative Anatomy.[2] He enrolled that winter in the private Anatomy classes of Dr. John Barclay and remained a pupil at his school on Surgeon's Square until 1824 when he taught the invertebrate part of Barclays' course (Anonymous, 1850:690; Anonymous, 1874-1875). He also continued to attend the lectures and courses of Robert Jameson, his earliest mentor, and played an active role in the Natural History Societies that had been established in Edinburgh. He was particularly active in the Wernerian Society of Natural History at Edinburgh University (a society established by Jameson in 1808) and was elected to the council of that society in 1826. He was also a member of the Plinian Society of Natural History also based at Edinburgh University. These societies gave him a forum for the presentation of his work and many of his early papers were published in the transactions of these societies. For the most part he presented papers on his work on marine invertebrates but he also gave papers on any interesting and unusual specimens which came his way, such as the long-nosed bandicoot (*Perameles nasuta*) and the swordfish (*Xiphias gladius*). He also published many papers in some of the newer journals at the University of Edinburgh including the *Edinburgh Journal of Science* and the *Edinburgh Philosophical*

Journal (later the *Edinburgh New Philosophical Journal*). These journals were particularly sympathetic to the controversial sciences coming from the continent of which Grant was an advocate.

Most of the papers that Grant published early in his career concerned the marine invertebrates or "zoophytes" that he collected around the coasts of Scotland and Ireland. He became particularly well known for the work he did on sponges. He definitively established that sponges were animals and not plants – a fact that had until then still been contested – and he also proved their porous nature by demonstrating that they had raised entrance and exit pores through which currents of water pass. In doing this he coined the group name Porifera which is still in use today. He named many new genera and species and several of those creatures still bear his names today. For example, he named the genera *Cliona* (1826), *Leuconia* (1826) and *Haliclona* (1826) as well as the species *Cliona celata* (1826), or yellow boring sponge, which bores into rocks and oyster shells. In addition, in 1828, in recognition of his contribution to their study, the zoologist and Unitarian minister John Fleming named the sponge genus *Grantia* after him (Fleming, 1828). By 1827 his work on sponges and other marine invertebrates was well known and had already been translated into French and German

(UCL special collections, Zoology, 1827 Applications; See also Desmond & Parker, 2006). It was in that year that he applied for the chair of Zoology at the University of London, recommended by his colleagues and friends at Edinburgh including Robert Jameson, David Brewster, John Fleming and Alexander Monro (tertius).

The University of London

Engraving of the University of London by T. H. Shepherd. Reproduced with permission. © University College London.

The newly founded University of London must have seemed the ideal location for a radical reformer such as Grant. It was founded along Benthamite principles, although Jeremy Bentham was not, one of the original founders as some historians have suggested,[3] nor was he part of the original council. The

University aimed not only to open up a University education to those who were denied it by the existing English Universities (Oxford and Cambridge were open only to members of the Church of England and only those that could afford the fees at that – Catholics, Jews and those without an affluent background, for example, were thus denied such an education and access to the professions that required it) but also to facilitate reform in the nation's institutions and professions. Its radical vision was to open up a liberal education to the middle classes. With the promise of a new, informed populous, it was hoped that these London graduates would force the political reforms which would transform the social, industrial and medical status quo (Desmond, 1989: 27). Grant, with his revolutionary science and passion for reform was, at this young institution, in a unique position to shape the course of the teaching and science of Zoology and Comparative Anatomy in England; to refresh a science which in England was stagnating and which was being far outstripped by its neighbours in Scotland and in Continental Europe.

Grant and his Museum

When he arrived to take up the chair, Grant found no teaching materials with which to conduct his lectures. He immediately set about establishing a museum to serve as the teaching collection for his Zoology and Comparative Anatomy courses. He amassed specimens for exhibition, materials for dissection, diagrams and lecture notes as well as making his personal library available for use by his students. Initially, the University provided moderate funds for furnishing this teaching collection with specimens, starting with the sum of £200, but over the years this fund gradually dried up until it all but stopped completely. In 1850 for example, Grant complains that the University had not bought any new specimens or dissection materials for 15 years.

For the most part Grant's "Zootomical Museum" was composed of specimens from his own collection, or materials that he had purchased for the benefit of the students from his own very meagre funds. The Museum and Grant personally, also received many donations from Grant's friends and colleagues, as well as in later years, students who had gone on to live and work all over the world. He received specimens from the London Zoological Gardens and the British Museum. Specimens were also donated by the other founding professors of the University of London, as well as Gideon Mantell and the zoologist Andrew Smith. Some of the better dissections by Grant's pupils were also placed in the museum for use in subsequent lectures. Many of the specimens that made up this original Museum can still be seen in the Grant Museum to this day (the Museum was re-titled in honour of its founder in 1997). By the 1850s the Museum consisted of about 10,000 specimens. It might have been more but the collection was constantly being depleted by rats and the wear and tear inflicted by students.

Grant used the specimens in the Museum for the demonstration of his courses and lectures. He believed that if one were to understand the more "complex" organisms, then one needed to understand the "simplest" first in order that one might understand their progression from simple to complex according to their common plan. He believed that one could trace the evolution of animals from the very simplest organisms right up to man. He also believed that we could see in the embryological development of individual organisms the stages through which those organisms passed in the course of their evolution. This view is known as *recapitulation*. These ideas guided the structure of the courses that Grant delivered as well as the organisation of his Museum.[4] His lectures were divided into sections, each dealing with a different system of organs of the body – for example the organs of respiration, circulation or reproduction – he would then discuss these systems for the different classes of the Animal Kingdom, starting with the most simple – the "animalcules" or the Protozoa – and working through to the most complex – the mammals. His Museum followed these principles as well, being organised so as to illustrate "the whole continuous chain of beings, from the lowest corals up to the highest animal forms that exist" (Grant, 1836c). Not all of the specimens that Grant had at his disposal were exhibited in his Zootomical Museum, those that were on display however, where possible, would contain several specimens in a single preparation, presumably dissected to show the various organs of the creature in question in a way that conveyed the maximum information to the students.

As well as having access to the specimens that composed the Museum itself, Grant's students also had the opportunity to undertake practical work of their own. He provided students, at his own expense, with materials and specimens for dissection work, as well as practical microscope instruction and the preparation of specimens for display and demonstration (UCL Special Collection MS ADD 59). It appears from comments made by Grant and his students in various published letters, that these dissections and practical sessions were done only with those students who expressed an interest in some further instruction, additional to Grant's weekly lectures. E. Ray Lankester (1847-1929), Grant's successor to the chairs of Zoology and Comparative Anatomy in 1875, is widely credited with having introduced practical laboratory work in a systematic

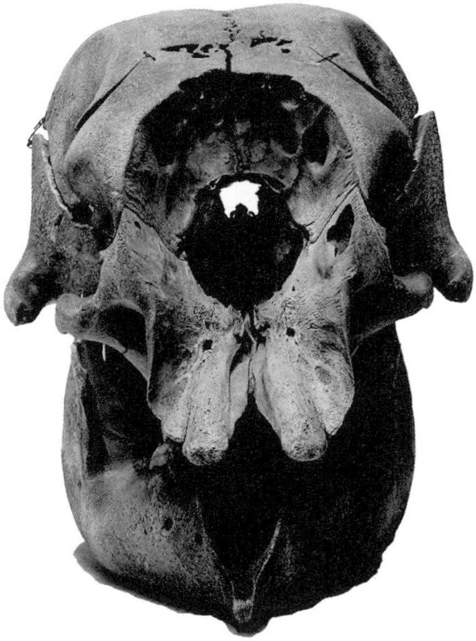

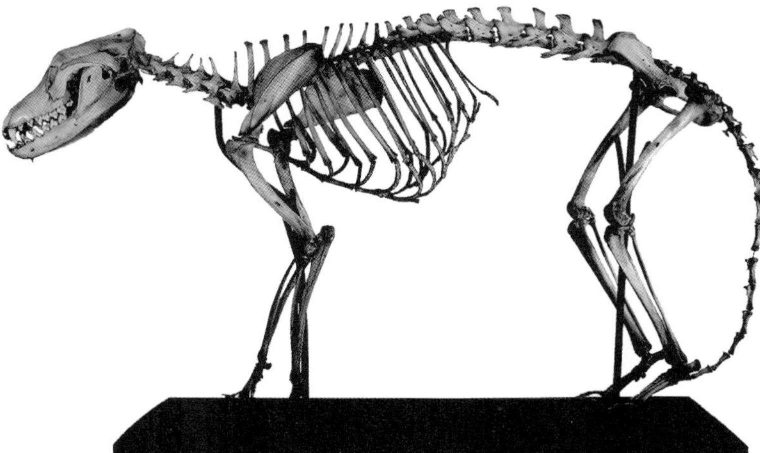

The juvenile elephant skull and the thylacine skeleton above are part of Grant's original teaching collection. These specimens can still be seen at the Grant Museum today. © The Grant Museum of Zoology, UCL.

and academically rigorous way into the department (Lester, 1995:65). Yet, we can see that Grant had already established such a practice among his own students, at least for those who wished to take their studies of Comparative Anatomy further. However, this is the same arrangement that Lankester came to with his students, albeit through an official agreement with the university and upon payment of a laboratory fee. We might suppose however, that by the time Lankester took up the post, Grant had largely ceased this practice, in part due to limitations of funds and in part due to his advanced age in the final years of his professorship – he lectured up until the month before his death and was 80 years old when he died. This would explain the contention that Lankester was the pioneer of this practice in academic Zoology and Comparative Anatomy classes. Though he may have introduced it on a scale never before seen, for Grant's earlier students at least, it was an integral part of their experiences of the study of the zoological sciences. He is certainly responsible for establishing Zoology and Comparative Anatomy as academic disciplines in Great Britain for the first time.

Grant and his students

It was not only through his ground-breaking lectures but also through his Museum and his commitment to his students that

Grant came to influence a generation of future zoologists, naturalists and medical men. He found time to encourage and advise his more diligent students as well as instructing them on the practical aspects of the discipline. He would often have favoured students over to his lodgings near Euston for breakfast where he would regale them with stories of his experiences in Scotland and on the Continent. Or he would discourse with them on subjects from "the power of independent life manifested by animal cells, or from some tale of the days of Waterloo, to descant upon the terpsichorean merits of the actress who then most commanded the attention of the town" (Poore, 1901). Other students he would invite on his frequent walking tours of the

Grant's house at 2 Euston Grove at the time of his death in 1874. These pictures were taken in 1962 just prior to the building's demolition. © The Grant Museum of Zoology, UCL.

continent, including, on more than one occasion, a young Indian student, Soorjo Coomar Goodeve Chuckerbutty.[5] These students he would take to the lectures, soirées and collections of some of the most eminent Continental comparative anatomists, affording these students a glimpse of what British Comparative Anatomy could be.

i. Charles Robert Darwin

Charles Darwin

Of all his students, Grant is undoubtedly most celebrated for the influence he had in Edinburgh on the young Charles Darwin. The relationship between Grant and Darwin has been subjected to minute scrutiny and so does not require detailed discussion here (Sloan, 1985; Hodge, 1985). Rather, it is worth adding to our picture of Grant the teacher by briefly considering the influence that he had over a number of other great scientists and medical men whom he taught after Darwin. However, it would be remiss not to make some mention of their relationship, thus a brief summary of their dealings with each other will be given before moving on to other matters.

Darwin became acquainted with Grant in 1826 while he was a second-year medical student at Edinburgh University. Darwin already had a considerable interest in Natural History and was enrolled in the lectures of Robert Jameson whose classes Grant had continued to attend after his return from Europe. Together Grant and Darwin searched for marine life in the Firth of Forth and Grant guided Darwin's study of them. Grant talked to Darwin about the work of Lamarck and of Darwin's own grand-father Erasmus. He also introduced Darwin to the transformist ideas that were to feature so prominently in his London lectures; that in order to understand the most complex organisms, one needs first to understand the simplest. Such an understanding, he believed would unlock the laws that govern all animals and their generation. It was these ideas which Darwin took with him on his *Beagle* voyages, though Darwin's own theories were soon to flourish quite independently and differently. Grant provided Darwin with copies of his early papers, works which Darwin later called "first-rate" (Barlow, 1958: 49). Together they worked on the generation of Grant's specialist subject, the zoophytes, and Darwin discovered that Grant's "law" concerning the free-swimming nature of their ova generalised to a number of other zoophytes, including the sea mat *Flustra*.[6] Grant also passed on some of his expertise with a microscope to Darwin and the two of them dissected the specimens they found on their expeditions together. They both attended meetings at the Plinian Society and Grant introduced Darwin to the Wernerian Society as his guest, presenting Darwin's findings concerning the *Flustra* along with his own.

Yet, in spite of their close relationship in Edinburgh, Grant and Darwin do not appear to have remained in contact when they both ultimately moved to London. The reasons for this are unclear and have been the subject of much speculation and theorising (Jesperson, 1949; Desmond, 1989; Gardiner, 1999), but it seems that a combination of factors may have contributed to this break-down in their relationship. Desmond (1989) speculates that initially there may have been some professional rivalry between the two men, both working as they did on the marine invertebrates and both being men with a keen sense of intellectual property. In addition, Darwin may have distanced himself from Grant due to differences in their political commitments and attitudes to reform; he may not have wanted his own ideas hi-jacked by the radical press and used as they used Grant's ideas. The difference in their social standing may also have been a contributing factor. Darwin was a gentleman naturalist while Grant had to work for his living, earning a pittance from his work as a professor, something that Darwin himself never understood. He said of Grant "...he published some first-rate zoological papers, but after coming to London as professor in University College, he did nothing more in science – a fact which has always been inexplicable to me" (Barlow, 1958:49). What Darwin failed to appreciate was that Grant was working flat out to make ends meet, delivering on average 200 lectures per academic year, and yet – since professors were paid according to the number of students that they could attract, and since Comparative Anatomy was not a compulsory subject for those enrolled on a medical course – he could earn less than £100 a year. Moreover, Grant was not idle in science either, while his publishing output certainly declined after his arrival in London and, in later years, his access to specimens was restricted, he continued to present his zoological findings to his classes and at his public lectures.[7]

While Darwin remains Grant's most high-profile student, there were many more students that Grant influenced in his lectures at the University of London. There are three in particular worth mentioning here: W. B. Carpenter (1813-1885), William Henry Flower (1831-1899) and Edwin Lankester (1814-1874). His influence over these men has not been as widely appreciated, and certainly the effects are less tangible than his impact on Darwin. Nonetheless, they require mention to illustrate the scope of Grant's influence. Those attending Grant's lectures at the

University of London varied from his colleagues at other institutions such as P. M. Roget (1779-1869); political men, such as Thomas Southwood Smith (1788-1861); to fresh-faced young men, leaving home for perhaps the first time. In particular however, his classes consisted of budding young medical men whose interests extended beyond those classes compulsory for attaining their medical degrees. Among this latter group was William Benjamin Carpenter.

ii. William Benjamin Carpenter

Carpenter was another medical man, like Grant, whose interests turned to the biological sciences. He came to the University of London as a student in 1835. Before that he had been apprenticed to a surgeon in Bristol and had travelled with him to the West Indies in 1833. On his return, he enrolled in Grant's lectures on Comparative Anatomy and these quickly had a strong effect on him. Already aware of the transformist theories of Geoffroy and Lamarck from his attendance of lectures at the Bristol Institution, Carpenter was a sympathetic and enthusiastic pupil of Grant's. In later years he remembered Grant's course fondly "not only for the information which he gained through it, but for the mental quickening and special love of the subject which it aroused in him" (Carpenter, 1888:10). His first paper O*n the Structure and Function of the Organs of Respiration*

(1835-1836) was written just after Carpenter took Grant's Comparative Anatomy course. In that paper he acknowledges the influence that Grant's lectures had on his work, as well as the influence of Carus, Roget, Tiedemann and Lindley. In particular, we can see how he defers to Grant's expertise on the structure of the sponge (1835-1836: 221) as well as to his own early mention of a unity of plan throughout the Animal Kingdom, indeed the whole of nature.

William Benjamin Carpenter. Reproduced with permission.
© National Portrait Gallery, London.

We can see in Carpenter's early work a considerable debt to Grant's lecture series at the University of London, a debt which he acknowledged all his life. Moreover, through the course of their lives Grant's and Carpenter's paths were to cross many times. After some time studying and working in Edinburgh and then Bristol, Carpenter came to settle in London. In 1845 he took up the lectureship in General Anatomy and Physiology at the London Hospital, and in 1847 he became the examiner in Physiology and Comparative Anatomy at the University of London, a position that would certainly have brought him into contact with Robert Grant. Two years later he took up the Professorship in Medical Jurisprudence at University College London, again, his contact with Grant would have been inevitable – particularly given that until this time Grant had been Dean of the Medical Faculty at University College. And their association would not just have been through the University of London. Carpenter held the Fullerian Professorship of Physiology (1844-1847) after Grant (1837-1840) at the Royal Institution of Great Britain, while Grant took over the Swiney Lectureship at the British Museum from Carpenter in 1853. Yet in spite of the professional contact that they must have had, little information exists regarding the extent of their friendship and communication after Carpenter's student days.

iii. William Henry Flower

Another of the students who found inspiration in the lectures of Robert Grant was William Henry Flower. In his day, Flower was a zoologist and museum curator, curating some of the most prestigious museums in London, among them the Hunterian

Museum at the Royal College of Surgeons (1861-1884) and the British Museum (Natural History) at South Kensington (1884-1898).[8] Even in his early life one can see the passion that he had for the structure and organisation of museums. As a child a room in his parents' house was given over to his collection of natural history objects and, while at school, he took over the curatorship of his school museum. Initially it seems from his notes that Flower's interest was primarily in birds. He filled notebooks with observations of sightings of birds and the dates on which the specimens he collected were shot and prepared.

In 1847, Flower came to study at University College London and later in his first year he became a student in Grant's Zoology class. We can immediately see Grant's influence on the young Flower in a letter that he wrote home to his mother.

> A knowledge of birds alone is of very little use; I want a general and scientific knowledge of the whole animal kingdom, particularly of the lower classes, mollusca and radiata, of which I know next to nothing, and without which I could never turn my love of natural history to any advantage (Quoted in Cornish (1904:7)).[9]

These words could have come straight from Grant's mouth and indeed echo the content of his inaugural lecture at the University of London (Grant, 1828a), a lecture he would certainly have delivered with little change to each successive class. Flower must have been a most attentive student, winning Grant's silver medal for Zoology. He also returned to Grant's classes towards the end of his time at University College in 1851, attending his lectures in Comparative Anatomy. He was beaten to the Gold medal, however, by one Joseph Lister (1827-1912), another of Grant's illustrious students.

William Henry Flower. Reproduced with permission.
© Natural History Museum, London.

Flower's time in Grant's classes, as well as his later experiences as curator of the surgical museum at Middlesex Hospital and lecturer on Comparative Anatomy will also have helped to shape Flower's broader views on the structure and organisation of a Zoology museum. These views he took with him to the

Hunterian Museum and the British Museum (Natural History). In his *Essays on Museums and other Subjects* (1898), Flower urges that Zoology and Comparative Anatomy should not be taught as separate subjects but should instead be combined. He also urged that in organising a collection to illustrate the Animal Kingdom, one should be equally sensitive to the combined nature of the subjects and, rather than separating out specimens to illustrate the various organs of the body, one should instead keep animals together, showing all aspects of their bodies and the stages of their development.

> In zoology the collection should consist of illustrations of the principal modifications of animal forms, living and extinct, a few selected typical examples of each being given, showing the anatomy and development as well as the external form (Flowers, 1889).

Much would have changed in the prevailing understanding of the Animal Kingdom and its evolution since Grant's day, not least given the publication of the *Origin of Species* (1859). However, it was from his earliest contact with the zoological sciences at University College that Flower would have appreciated the importance of the combined approach to Zoology and Comparative Anatomy. He entered a department with the first ever combined chairs of these two subjects, taught with an extensive museum of specimens. Moreover, these specimens would have been arranged just as Flower urged; the recent and fossil specimens organised not according to the different functional parts of their bodies – the different systems grouped together, with the same animals thus displayed in different locations – but according to their place in the great chain of being as Grant perceived it. However, while Grant's teaching and his methods may have had this early effect on Flower, it is undoubtedly the case that Flower took museum organisation to a whole new level. Where Grant was concerned simply with the illustration of his lectures, Flower wanted to accommodate both students of Natural History and members of the general public, making the museum displays accessible to both.

Edwin Lankester. Reproduced with Permission. © National Portrait Gallery, London.

iv. Edwin Lankester

The final student to be discussed is Edwin Lankester. Lankester was a contemporary of Carpenter's and the two men remained in contact over the years, Lankester writing the preface for the second edition of Carpenter's *Vegetable Physiology and Systematic Botany* (1858). Many years before however, Lankester had enrolled at the

University of London intent on gaining a medical qualification. He had already spent five years as an apprentice and then assistant to surgeons in his home town of Woodbridge in Suffolk as well as in Hampshire, Essex and London. However, in 1834 he was provided, by friends, with the funds to attend University College. While there his passions leant more towards Botany than to Zoology, and yet it is evident that he held his old Zoology professor in high esteem. Years later, in 1853, in recognition of his years spent teaching and researching at University College, Grant's friends and colleagues got up a public testimonial to raise funds to provide him with an annuity. They raised £50 for a magnificent microscope and presented him with an annuity of 50 guineas a year.[10] The microscope was engraved with the following words:

> Presented to
> ROBERT EDMOND GRANT, M.D., F.R.S.L. & E., F.L.S., F.G.S., &c.,
> Professor of Comparative Anatomy and Zoology in University College, London, by his friends and former pupils,
> AS A TESTIMONY
> Of their esteem for his private worth, and of their sense of his eminent services in the cause of science.
> 1853.

Lankester was among those friends and former pupils who presented the microscope to Grant. The choice of gift, at least for Lankester would also have been significant. Lankester went on to become the president of the Microscopical Society and co-founder of the *Quarterly Journal of Microscopical Science*, as well as fellow of the Royal Society. It would have been in Grant's classes, as well as those of Lankester's mentor the Botany professor John Lindley (1799-1865) that Lankester would first have really engaged in microscope work.

The Microscope presented to Grant by his friends and colleagues in 1853.

As well as their common passion for the microscope, Lankester and Grant were both also ardent medical and social reformers and both counted Thomas Wakley among their close acquaintances. They were both also acquainted with Thomas Henry Huxley (1825-1895), Lankester was particularly close to him and Huxley would often spend time with the Lankester family, and was instrumental in the later career of Edwin Lankester's oldest son E. Ray Lankester. It was E. Ray Lankester who took over the chairs of Zoology and Comparative Anatomy in 1875 after Grant's death the year before.

Grant and the Medical Profession

While much of Grant's life and work was devoted to the study and teaching of Comparative Anatomy he did not altogether abandon the profession to which he had first been called. At Edinburgh he delivered a number of papers to the Royal Medical Society of Edinburgh of which he was a member and also joined a number of the other student medical societies. Later, in London he continued his connection with the medical profession. The subject of Comparative Anatomy, while it was not compulsory for those undertaking a medical degree at the University of London, was nonetheless recommended and most of those attending Grant's university lectures were future medical men. Grant was also Dean of the Medical Faculty at UCL from 1847 to 1849, and was active in the medical reform movements of the day, giving inflammatory speeches at the British Medical Association of which he was one of the earliest members (Grant, 1841b). Moreover, Grant's transformist Comparative Anatomy was championed by the radical medical journals, specifically the *Lancet*. However, despite his connections with prominent medical men of the day and his role in medical education at UCL, Grant probably never practiced medicine in England.

At that time the Royal College of Physicians held the monopoly on licenses to practice medicine in England. Those medical men who had received their medical training and licenses outside England were obliged to take a further examination in order that they might receive a license to practice in London. As a graduate of Edinburgh University, a fellow of the Royal College of Physicians in Edinburgh and a Scottish license holder Grant found the practice not just insulting but fraudulent, "arbitrary, illegal and ignominious" (Grant, 1841b) and he refused to submit to such an indignity. As a result, Grant was unable to supplement his income with medical practice as many of his other colleagues in London did. There were times when he earned less than £100 per year from his fees and an additional income would have gone a long way to alleviating his financial hardship. Contrast this with William Sharpey's £686 per year from the University of

London chair of Physiology, or Marshall Hall's earnings in the region of £2,000 per year from his practice and we can see just how impoverished Grant was in comparison with some of his colleagues (Desmond, 1984).

However, Grant was not always prohibited from practicing medicine. With the passing of the 1858 medical act, the medical degrees of Edinburgh University were finally recognised in England without the need for an additional examination through the Royal Colleges of Physicians and Surgeons and Grant registered with the General Council of Medical Education and Registration in 1859.[11] However, at the age of 66, having not practiced medicine for 39 years (if indeed he ever had), as well as still teaching approaching 200 lectures a year, it is unlikely that Grant actually exercised his entitlement, certainly, no known record exists of his having done so. Thus, even though he no longer needed to take his stand against the Royal College of Physicians his financial hardships were no further eased. Nevertheless, we might view his eventual registration as a triumphant conclusion to his years of principled self-sacrifice.

Grant's decline

Grant's professional reputation reached its peak in the late 1820s and early 1830s; his publishing output was at a career high and his lectures on Comparative Anatomy published in the *Lancet* were hailed as the state of the art. He was on the councils of the Zoological Society and Geological Society of London, a fellow of the Royal Society, in 1837 Fullerian Professor of Physiology at the Royal Institution of Great Britain and in 1853 Swiney lecturer at the British Museum. He was beloved of the radical press and welcomed in the universities and museums of Europe. Yet his success was short-lived and his professional life began a steady decline. However, this decline was not as steep as Darwin's recollections would have us believe; he did not simply cease to work in science on taking up his chair at University College London. The source of Grant's decline has been treated extensively elsewhere (Desmond, 1984; 1989) but seems to have stemmed from a number of personal, political and professional factors.

First, Grant was unable to supplement his income as a professor with medical practice. The only funds that he had available to him were those he could raise from the teaching he undertook. Initially, the University of London professors were guaranteed £300 for their services, but by 1831, they were expected to survive on their fees alone; Grant's barely averaged £100 per year. In the session 1842-1843 Grant earned £135.10, while in the following session he took home just £94 (figures taken from

Desmond, 1984: 143; see also the student registers held in the University College London records office).

Grant was able to alleviate his financial hardship somewhat. He gave public lectures at private schools and scientific societies for a modest fee and there were also a number of paying positions available in London in the short-term, including the Fullerian Professorship and Swiney lectureships already mentioned. However, there was stiff competition for these positions and opportunities were few and far between. This meant, however, that Grant was constantly engaged in some teaching activity or another, whether it was at the University or elsewhere. And if he was not lecturing then he was preparing for the lectures, or engaged in some dissection for use in the illustration of his courses. He thus had little time for undertaking his own research unlike in his Edinburgh days. And, as time went on, and Grant became gradually older and deafer, the heavy lecture load must have taken its toll, in spite of his continuing “vigorous strides” (Poore, 1901:190).

Grant’s poverty was eased somewhat in later years. In 1850 the University presented him with a stipend of £100 to help bolster his flagging student numbers. He was also helped by the annuity of 50 guineas which his friends and colleagues presented him with in 1853. In addition, he was left a small sum on the death of one of his brothers. Yet, it did not free him from the need to teach, or even help to reduce his always excessive workload. Grant was thus never able to devote himself purely to research, a luxury which many of his contemporaries enjoyed.

Second, Grant's controversial evolutionary ideas, radical politics and uncompromising position with respect to reform meant that he did much to alienate the conservative elite of the scientific societies. His calls for reform were greeted with an unfavourable reception and he eventually withdrew from active membership of these societies; most notably in the case of the Zoological Society of London and the Royal Society. At the Zoological Society of London Grant had been the object of some acclaim. He gave a series of lectures to the members in 1833 (Grant, 1833j), and presented many individual papers before the Society (Grant 1833b-i). He was also elected to the council and sat on several of the Society's committees. However, there was a large Tory contingent among the councillors of the Society, among them the Zoologist Richard Owen (1804-1892). Grant and Owen had enjoyed an amiable acquaintance in Paris back in 1831. However, this friendship took an unpleasant turn back in London. Owen's views on the structure of the Animal Kingdom were significantly opposed to Grant's. He firmly rejected the

Geoffroyan unity of plan and the transmutationist theories of Lamarck which Grant championed. Grant's increasing influence and the popularity of his science at the Society would have been unsettling to him. The two men also found themselves in competition for specimens at the Zoological Society and access to its menagerie, for they were the only comparative anatomists to publish extensively on those specimens (Desmond, 1989:140-141). At the same time as this acrimonious rivalry, there was a radical movement within the Society for reform.

These reformers wanted to see a more democratic structure to the management of the society, a reform in its financial priorities and more accountability to the membership. This uprising was not to be tolerated and Grant, one of the leading members of the reformers was voted off the Zoological Society council amid a calamitous furore. Disgusted by the underhanded dealings of the conservatives Grant withdrew from the Society and in so doing lost access to specimens for dissection and study and an important forum for the presentation of his work. This was another example of Grant's principled stand thwarting his professional ambition and was certainly a contributing factor to his later decline. His conflict with Owen, however, was far from resolved and the two continued to wrangle publicly and through their respective institutions (Owen was curator of the Hunterian

Museum at the Royal College of Surgeons), at one point fighting an increasingly bitter battle for possession of a precious cargo of specimens from New Zealand which had been mistakenly delivered to Owen instead of Grant, a delivery that Owen was loath to give up.

At the Royal Society we see a similar story, this time, however, laced with accusations of plagiarism and personal recriminations. Against a background of societal unrest Grant made two accusations of plagiarism in the 1830s against fellows of the society. The first accusation was against George Newport, who won a Royal Society medal for work that had been pirated, Grant contended, from his own public lectures of 1832. The second accusation was levelled against Peter Mark Roget (1779-1869), the secretary of the Royal Society, whose Bridgewater Treaties was widely acknowledged in the radical press as plagiarized from the same lectures as those attended by Mr. Newport (Grant 1846a, b).[12] Newport and Roget's response was to contend that as a "public professor" the content of Grant's lectures was the "property of those to whom they are communicated", they believed that they were thus at liberty to make free with the content of those lectures as they saw fit (Roget, 1846). The question over intellectual property was impossible to resolve. Whatever we might conclude about the veracity of these claims though, it does show that Grant presented his scientific discoveries as part of his lectures rather than through their publication. Indeed, many of the published papers that present Grant's ideas were, in fact, reports of his discoveries by third parties, rather than papers in Grant's own hand (see for example, Anonymous 1825a, 1825b).[13]

Newport's medal was not the only Royal Society Medal that gave Grant cause for concern. In 1847 along with Thomas Wharton Jones, James Copland, Marshall Hall, William Frederick Chambers, James Ranald Martin, John Webster, Richard Bright, George Gulliver, and Robert Lee, Grant signed a requisition requiring a Special General Meeting to determine the legality of the 1845 Royal Medal in Physiology awarded to a Mr. Beck. However, the real issue here was not so much the legality of the medal presented as the power struggle between the conservative and radical fellows of the society. Grant's disgust with the secretive cliques of the Royal Society grew and in 1850 he refused the opportunity to join the Royal Society's committee of Zoology and Animal Physiology (Grant 1850). Once again, through his unwavering commitment to his ideals, and in part in protest at his own and his colleagues' shabby treatment at the hands of the Society, Grant shut another door against potential sources of professional influence and support.

Finally, in his day, Grant's views were considered dangerously controversial by some, and cutting edge by others. However, he continued to cling to an outdated comparative anatomy in his later years. While he acknowledged the work of Darwin and other contemporary evolutionists, dedicating his final work to the former (Grant, 1861), Grant continued to teach the works of Geoffroy and Lamarck and to preach spontaneous generation and transmutation. He was thus left behind as the new evolutionary sciences swept past him, much to the frustration, and, at times, secret amusement of his students. It took E. Ray Lankester's arrival in the chair to bring the science of Comparative Anatomy up to date again at University College London.

Yet, in spite of this decline, it would be a mistake to overlook the significant influence that Grant had in his day. His published output may have dwindled, but his last work was published in 1861, when Grant was 68 years-old, and though his transmutationist framework may have been outdated in that later work, those who took up his mantel as teachers of Comparative Anatomy owe much to a man who is now frequently remembered as little more than an old-fashioned eccentric in funny clothes (Poore, 1901; Shäfer, 1901; for more modern portrayals of Grant in this light see Harte & North, 2004).

Grant as an old man. © The Grant Museum of Zoology, UCL.

An enduring legacy

On 22nd August 1874 William Sharpey, professor of Anatomy and Physiology, reported to Mr. J. Robson, the Secretary of University College, that Professor Grant had been taken ill with dysentery some days before. When he had gone to visit him Sharpey had found him weak but defiant, refusing to accept that he might be dying, and talking of his specimens at the College. Sharpey asked him what he would wish to do with his specimens if he should pass away and Grant stated that he would wish them to go to the College. Sharpey, pressing further, asked whether Grant had any family or friends that he would want to benefit from his estate on the event of his death. Grant however, said that he did not; all his brothers and sisters having died before him without children. Mindful that all Grant's property would go to the exchequer should he die intestate, Sharpey suggested that Grant leave his money, library and personal effects to the College for the purpose of maintaining and updating the Museum. Grant accepted the suggestion enthusiastically and a paper was drawn up giving the details, which Grant duly signed "with difficulty – but hand unsupported". The following day however, Grant became distressed and disorientated, believing that he had signed everything he owned and had worked for his whole life away to the University. Sharpey comforted him, reminding him that it would only go to the College in the event of his death. Grant deteriorated rapidly from then on, his speech became indistinct and his ability to give "intelligible and connected instructions" was diminished. He died on 23rd August 1874 in the very year he had intended to retire and was buried in Highgate Cemetery.

Grant's library, specimens and what money he had came to the University as he wished and formed the foundation of the Zoology library as well as furnishing the library with many more much needed books. His Museum continued to be used as it was in his day and was expanded and updated in accordance with his will over the successive curatorships. E. Ray Lankester, Walter Weldon (1860-1906), J. P. Hill (1873-1954) and D.M.S Watson (1886-1973) all held Grant's chair in turn, bringing to the collection specimens which reflected their own particular interests. There remains a mystery relating to Grant's legacy however, the whereabouts of his personal letters, notebooks and journals.

Grant's private papers

While Grant's specimens and library all came to the University on the execution of his will his personal papers did not. We know from the biographical sketch that Grant kept bound volumes of his letters, detailed notebooks of his zoological researches, and journals.

> Early impressed at college with the importance of taking and preserving notes of observations, thoughts, readings, he has preserved that habit with great constancy, for the last forty years; and he still continues, in his summer annual visits to the continental museums of natural history and anatomy, to keep regular journals of his observations. Manuscript copies also of private letters, written at home or abroad, are preserved in volumes; and even the subject and date of every lecture of the thousands he has delivered for the past quarter of a century are methodically registered and retained (Anonymous, 1850: 692).

However, we also know from that same source that Grant did not keep copies of the manuscripts for his publications, these he burned. This has led some researchers, including Desmond (1984) to suggest that Grant burned these personal papers before his death. However, such a fate is unlikely when we consider a number of factors.

First, as we have seen from Sharpey's account of Grant's death, he did not believe that he was actually dying and was concerned that he would not be able to go back to work once he recovered, having given over all his worldly goods to the University. It seems unlikely therefore that he would take such a final step with his own personal papers. Second, we can also see that by the time it became evident that Grant was dying, he was too weak to move from his room. If we consider what would have been involved in burning these documents, it becomes increasingly implausible that he would, or indeed could, have burned them. We know from the biographical sketch that he kept these bound volumes of letters, journals and notebooks for forty years by 1850, if he continued in this practice until his death then that totals 64 years of letters and journals and so forth. If we suggest that he had one volume of letters for each year, one notebook and one journal then we have 192 volumes to be disposed of in a Victorian townhouse bedroom fireplace. He would also have had to rip these volumes into pieces in order to get them into the fireplace. That an 80 year-old man, confused, distressed and dying of dysentery, could have performed such a feat seems all but impossible. This suggests the need for an alternative explanation.

There are a few possible candidates. He may, for example, have destroyed them sometime before he became ill. However, given his proprietorial attitude to his discoveries and the pride he had in his work and associations, it is hard to see what motivation he might have had at that time. On the other hand, he might have given them to a friend or colleague to destroy. Again we must ask why. Desmond has suggested that there might be something in the rumours that Grant was homosexual.[14] However, there is little in the contemporary accounts of Grant's life other than his bachelor's existence which supports this. Yet, even supposing that it is true, it is unlikely that any personal discussion of his sexual orientation would have featured in his scientific notebooks, or that any of the more delicate correspondence would have been preserved in the volumes. However, if they were, we must concede that a close friend might have been entrusted with the task of destroying them. What is more likely is that Grant did indeed give his letters to one of his old friends, but rather in the spirit of continuing his researches, than hiding any traces of his personal life. It might even be the case that the letters found their way into the hands of one of the scrap merchants of the day. We just don't know. If these letters did come to light however, they would contain some of the most exciting correspondence with some of the foremost thinkers of the day including Cuvier, Geoffroy, Henri de Blainville (1777-

1850), Jameson, Darwin and Huxley, and they would shed light on some of the more controversial and less documented aspects of Grant's life.

The Museum Today

Grant's Museum has moved many times since Grant's day and, although the collection has increased considerably in size – now numbering some 33,000 specimens – the space accommodating the Museum itself has gradually decreased. From its original location in the South Cloisters in the Wilkins Building, the Museum is now housed in the Darwin Building further down

The Grand Museum of Zoology and Comparative Anatomy as it looks today. © The Grant Museum of Zoololgy, UCL.

Gower Street. The Museum now contains not only those specimens bequeathed to UCL by Grant and those amassed by subsequent curators, but also the collections of Thomas Henry Huxley, the laryngologist Sir Victor Negus, and a beautiful collection of Blaschka glass models. The Museum is still used as it was in the 19th Century in the illustration of Zoology practical classes and lectures. However, its uses now go beyond those originally envisaged by Grant. The collection is used by artists and Museum Studies students, by school groups and researchers and is also open to the public. Today the Museum reaches more people than ever before.

Bibliography of Robert Grant's Published Work

An annotated bibliography of Grant's published works can be found in Desmond & Parker (2006) consisting of 128 works.

1814. *Dissertatio physiologica inauguralis, de circuitu sanguinis in foetu*. Edinburgh: Ballantyne. Pp 28.

1825

a) Observations and experiments on the structure and functions of the sponge. *Edinburgh Philosophical Journal* 13: 94-107, 333-346.

b) Dr. Grant on the ova of the sponge. *Edinburgh Philosophical*

Journal 13: 381-383.

1826

a) Observations and experiments on the structure and functions of the sponge. *Edinburgh Philosophical Journal* 14: 113-124, 336-341.

b) An appearance seen on the surface of the living *Corallina officinalis*. *Edinburgh Philosophical Journal* 14: 183.

c) On the spicula of the *Spongilla friabilis*, Lamarck. *Edinburgh Philosophical Journal* 14: 183-185.

d) Sounds produced under water by the *Tritonia arborescens*. *Edinburgh Philosophical Journal* 14: 185-186.

e) On the structure and nature of the *Spongilla friabilis*. *Edinburgh Philosophical Journal* 14: 270-284.

f) Notice of a new zoophyte (*Cliona celata*, Gr.) from the Firth of Forth. *Edinburgh New Philosophical Journal* 1: 78-81.

g) Observations on the spontaneous motions of the ova of the *Campanularia dichotoma*, *Gorgonia verrucosa*, *Caryophyllea carycularis*, *Spongia panicea*, *Sp. pappilaris*, *cristata*, *tomentosa*, and *Plumularia falcata*. *Edinburgh New Philosophical Journal* 1: 150-156.

h) Remarks on the structure of some calcareous sponges. *Edinburgh New Philosophical Journal* 1: 166-171.

i) On the siliceous spicula of two zoophytes from Shetland.

Edinburgh New Philosophical Journal 1: 195-196.

j) Observations on the structure of some silicious sponges. *Edinburgh New Philosophical Journal* 1: 341-351.

1827

a) Observations on the structure and functions of the sponge. *Edinburgh New Philosophical Journal* 2: 121-141.

b) Notice of two new species of British sponges. *Edinburgh New Philosophical Journal* 2: 203-204.

c) On the structure and characters of the *Octopus ventricosus*, Gr. (*Sepia octopodia*, Pent.), a rare species of octopus from the Firth of Forth. *Edinburgh New Philosophical Journal* 2: 309-317.

d) Observations on the structure and nature of *Flustrae*. *Edinburgh New Philosophical Journal* 3: 107-118, 337-342.

e) On the existence and uses of ciliae in the young of the gasteropodous Mollusca, and on the causes of the spiral turn of univalve shells. *Edinburgh Journal of Science* 7: 121-125.

f) On the structure and characters of the *Lernaea elongata*, Gr. a new species from the Arctic seas. *Edinburgh Journal of Science* 7: 141-154.

g) Notice regarding the ova of the *Pontobdella muricata*, Lam. *Edinburgh Journal of Science* 7: 160-161.

h) Notice regarding the structure and mode of generation of the

Virgularia mirabilis and *Pennatula phosphorea. Edinburgh Journal of Science* 7: 330-334.

1828.

a) *An Essay on the Study of the Animal Kingdom. Being an Introductory Lecture Delivered in the University of London, on the 23rd of October, 1828*. London: Taylor. Pp 35

b) Observations on the generation of the *Lobularia digitata*, Lam. (*Alcyonium lobatum*, Pall.). *Edinburgh Journal of Science* 8: 104-110.

c) On the viscera of the common swordfish (*Xiphias gladius*, Lin.). *Edinburgh Medico-Chirurgical Society Transactions* 3: 79-93.

1829.

a) On the influence of light on the motions of infusoria. *Edinburgh Journal of Science* 10: 346-349.

b) Further observations of the generation of the *Virgularia mirabilis*. *Edinburgh Journal of Science* 10: 350-351.

c) [Letter to E. Geoffroy St Hilaire], in GEOFFROY St HILAIRE, E. Considérations sur des oeufs d'ornithorinque, Formant de nouveaux documens pour la question de la classification des monotrèmes. *Annales des Sciences Naturelles* 18: 157-164

1830.

a) Zoophytology. *Edinburgh Encyclopaedia* 18(2): 838-846.

b) Baron Cuvier. *Foreign Review and Continental Miscellany* 5: 342-380.

c) On the egg of *Ornithorynchus*. *Edinburgh New Philosophical Journal* 8: 149-151.

1831.

a) Extracts from a correspondence on the *Filaria medinensis* among some of the medical officers in the Honourable East India Company's service at Bombay; with a letter from Dr Robert Grant, Professor of Comparative Anatomy in the University of London. *Edinburgh Medical and Surgical Journal* 35: 112-118.

b) Long-Nosed *Perameles*, or pouched badger of New Holland. *Perameles nasuta* [re-printed letter accompanying plate X], in WILSON, J. *Illustrations of Zoology: being Representations of New, Rare, or Remarkable Subjects of the Animal Kingdom, Drawn and Coloured after Nature, with Historical and Descriptive Details*. Edinburgh: Blackwood.

1832.

a) On the structure of the eye of the swordfish (*Xiphias gladius*, Lin.). *Memoirs of the Wernerian Natural History Society* 6: 113-122.

b) Observations on the anatomy of the paca of Brazil,

(*Coelogenus*, F.Cuv.). *Memoirs of the Wernerian Natural History Society* 6: 133-143.

c) Observations on the anatomy of the *Perameles nasuta*, from New Holland. *Memoirs of the Wernerian Natural History Society* 6: 184-202.

1833.

a) *On the Study of Medicine: being an Introductory Address Delivered at the Opening of the Medical School of the University of London, October 1st, 1833*. London: Taylor. Pp 20.

b) On the nervous system of *Beroë pileus*, Lam., and on the structure of its cilia. *Proceedings of the Zoological Society of London* 1: 8-9.

c) On *Ianthina vulgaris*, Lam., and *Velella limbosa*, Lam. *Proceedings of the Zoological Society of London* 1: 14.

d) On the zoological characters of the genus *Loligopsis*, Lam., and account of a new species from the Indian Ocean. *Proceedings of the Zoological Society of London* 1: 26-27.

e) On a new species of *Sepiola* (*Sep. stenodactyla*) from the Mauritius, presented by C. Telfair, Esq. *Proceedings of the Zoological Society of London* 1: 42-43.

f) On the structure of the heart and the distribution of the blood-vessels of the large Indian tortoise (*Testudo Indica* [sic], Linn.) *Proceedings of the Zoological Society of London* 1: 43-44.

g) On the cranium of the round-headed grampus (*Delphinus globiceps*, Cuv.) *Proceedings of the Zoological Society of London* 1: 65-66.

h) On the cloaca of the female condor (*Sacorhamphus gryphus*, Dum.) *Proceedings of the Zoological Society of London* 1: 78.

i) On the Anatomy of *Loligopsis guttata*, Grant, and *Sepiola vulgaris*, Leach. *Proceedings of the Zoological Society of London* 1: 90-91.

j) *Outline of a Course of Lectures on the Structure and Classification of Animals, to be Delivered to the Members of the Zoological Society of London, in their Museum, to Commence on Tuesday the 15th of January, 1833, and to Continue on the Succeeding Tuesdays and Thursdays, at Half-past Seven o'clock p.m.* London: Mills, Jowett, and Mills. Pp 26.

1833-4.

Lectures on comparative anatomy and animal physiology, delivered during the session 1833-4. *The Lancet* 21-22.[15]

Lecture I. On the history and present state of the science.

Lecture II. On the objects of comparative anatomy and its applications to other branches of knowledge.

Lecture III. On the classification of the Animal Kingdom.

Lecture IV. On the classification of the organs of animals, and on the organs of support in animalcules and poripherous animals.

Lecture V. On the organs of support of polypiferous animals.

Lecture VI. On the organs of support of Acalepha and Echinoderma.

Lecture VII. On the organs of support of the helminithoid classes of articulated animals.

Lecture VIII. On the organs of support of the articulated classes of animals, provided with articulated members; viz Myriapoda, Insecta, Arachnida, and Crustacea.

Lecture IX. On the organs of support of the tunicated and conchiferous classes of molluscous animals.

Lecture X. On the organs of support of the gasteropodous and pteropodous Mollusca.

Lecture XI. On the organs of support of the cephalopodous Mollusca.

Lecture XII. On the development of the skeleton of the vertebrated classes, and the osteology of fishes.

Lecture XIII. On the osteology of fishes.

Lecture XIV. On the osteology of amphibious animals or Batrachia.

Lecture XV. On the osteology of serpents.

Lecture XVI. On the osteology of saurian reptiles.

Lecture XVII. On the osteology of chelonian reptiles.

Lecture XVIII.On the osteology of birds.

Lecture XIX. On the osteology of birds.

Lecture XX On the osteology of cetaceous Mammalia.

Lecture XXI. On the osteology of ruminating quadrupeds.

Lecture XXII. On the osteology of Pachyderma.

Lecture XXIII. On the osteology of Pachyderma.

Lecture XXIV. On the osteology of the monotrematous and edentulous quadrupeds.

Lecture XXV. On the osteology of Rodentia and Marsupialia.

Lecture XXVI. On the osteology of carnivorous quadrupeds.

Lecture XXVII. On the osteology of Insectivora and Cheiroptera.

Lecture XXVIII. On the osteology of Quadrumana and Bimana.

Lecture XXIX. On the ligaments and articulations, or the organs of connexion.

Lecture XXX. On the organs of motion in the cyclo-neurose or radiated classes of animals.

Lecture XXXI. On the muscular system of the articulated classes.

Lecture XXXII. On the muscular system of the molluscous classes.

Lecture XXXIII. On the muscular system of fishes and amphibians.

Lecture XXXIV. On the muscular system of reptiles and birds.

Lecture XXXV. On the muscular system of Mammalia.

Lecture XXXVI. On the nervous system of the radiated and helminthoid classes.

Lecture XXXVII. On the nervous system of the entomoid and molluscous classes.

Lecture XXXVIII. On the nervous system of the Vertebrata.

Lecture XXXIX. On the organs of vision.

Lecture XL. On the organs of hearing, smell, taste and touch.

Lecture XLI. On the digestive organs of the radiated or cyclo-neurose animals.

Lecture XLII. On the digestive organs of the articulated or diploneurose classes.

Lecture XLIII. On the digestive organs of the molluscous or cyclo-gangliated classes.

Lecture XLIV. On the digestive organs of fishes, Amphibia, and reptiles.

Lecture XLV. On the digestive organs of birds and Mammalia.

Lecture XLVI. On the chyliferous system in the vertebrated classes.

Lecture XLVII. On the sanguiferous system in the radiated and articulated classes.

Lecture XLVIII. On the sanguiferous system in the molluscous classes.

Lecture XLIX. On the sanguiferous system in the vertebrated classes.

Lecture L. On the respiratory organs of the invertebrated animals.

Lecture LI. On the respiratory organs of the vertebrate.

Lecture LII. On the secreting organs of animals.

Lecture LIII. On the secreting and lymphatic systems in animals.

Lecture LIV. On the excreting and tegumentary organs.

Lecture LV. On the generative system in the radiated or cyclo-neurose classes.

Lecture LVI. On the generative system in the articulated classes.

Lecture LVII. On the generative system in the molluscous

classes.

Lecture LVIII. On the generative system of fishes and amphibian.

Lecture LIX. On the generative system of reptiles and birds

Lecture LX. On the generative system of Mammalia.

1834. On a fossil tooth found in a red sandstone above the coal formation in Berwickshire. *Edinburgh New Philosophical Journal* 16: 38-43.

1835.

a) On the nervous system of *Beroë pileus*, Lam., and on the structure of its cilia. *Transactions of the Zoological Society of London* 1: 9-12.

b) On the structure and characters of *Loligopsis*, and account of a new species (*Lol. guttata*, Grant) from the Indian seas. *Transactions of the Zoological Society of London* 1: 21-28.

c) On the anatomy of the *Sepiola vulgaris*, Leach, and an account of a new species (*Sep. stenodactyla*, Grant) from the coast of Mauritius. *Transactions of the Zoological Society of London* 1: 77-86.

1836.

a) Animal kingdom, volume 1, pp 107-118 in TODD, R. B. (editor), *The Cyclopaedia of Anatomy and Physiology*. London: Sherwood, Gilbert & Piper.

b) Chyliferous system, volume 1, pp 600-601 in TODD, R. B. (editor), *The Cyclopaedia of Anatomy and Physiology*. London: Sherwood, Gilbert & Piper.

c) [Evidence before committee] Report of the select committee appointed to inquire into the condition, management and affairs of the British Museum (Parliamentary papers, 14 July 1836), X: 19-31, 84-95, 124-37.

1838.

a) *On the Principles of Classification, as Applied to the Primary Divisions of the Animal Kingdom*, pp 262-310 in THOMSON, R. D. (editor), *The British Annual, and Epitome of Science for 1838*. London: Baillière.

b) Reply to Mr. Newport's insinuations respecting the writings of Dr. Marshall Hall and Dr. Grant. *The Lancet* 29: 746-748.

c) Further observations on Dr. Hall's statement regarding motor nerves of the articulata. *The Lancet* 29: 897-900.

d) Antediluvian remains at Stourton Quarry. *Liverpool Mercury*, 24 August 1838. Extracted as On the impressions of footsteps of *Chirotherium* in the Stourton Quarries at Liverpool. *Magazine of Natural History* 3: 43-48.

1839.

a) *General View of the Characters and the Distribution of Extinct Animals*, pp 222-281 in THOMSON, R. D. (editor), *The*

British Annual, and Epitome of Science for 1839. London: Baillière.

b) Digestive canal, pp 27-30 in TODD, R. B. (editor), *The Cyclopaedia of Anatomy and Physiology*. London: Sherwood, Gilbert & Piper.

1841

a) *Outlines of Comparative Anatomy, Presenting a Sketch of the Present State of Knowledge, and of the Progress of Discovery, in that Science; and Designed to Serve as an Introduction to Animal Physiology, and to the Principles of Classification in Zoology*. London: Baillière. Pp 656.

b) *On the Present State of the Medical Profession in England; being the Annual Oration Delivered before the Members of the British Medical Association, on 21st October 1841.* London: Renshaw. Pp 98.

1842. On the structure and history of the mastodontoid animals of North America. *Proceedings of the Geological Society* 3: 770-771.

1845

a) On the structure and history of polygastric animalcules. *Transactions of the British and Foreign Institute* 1: 353-358.

b) Lecture on the structure and history of the zoophytes. *Transactions of the British and Foreign Institute* 1: 378-380.

1846

a) Dr. Roget's Bridgewater treatise. *The Lancet* 47: 445-446.

b) [Letter], pp 482-483 in ROGET P. M., Dr. Roget's Bridgewater treatise. *The Lancet* 47: 482-483.

1849. [Letter to Charles J. B. Williams, 24 March 1849], p. 244 in WILLIAMS, C. J. B. *Memoirs of Life and Work.* London: Smith, Elder, & Co. Pp 522.

1850. [Letter to Thomas Bell informing the Royal Society of his refusal to join the committee of Zoology and Animal Physiology]. *The Lancet* 55: 88.

1861. *Tabular View of the Primary Divisions of the Animal Kingdom, Intended to Serve as an Outline of an Elementary Course of Recent Zoology (Cainozoology), or the Nature of Existing Animals*. London: Walton & Maberly. Pp 91.

Notes

[1] A writer to the signet was the Scottish equivalent of a solicitor.
[2] On returning to Edinburgh Grant probably began to practice medicine in addition to pursuing his interest in Natural History. (Anonymous, 1850: 690) However, another source suggest that he did not (Anonymous, 1874-1875). The former source was written with Grant's co-operation during his lifetime and is thus more likely to be an accurate reflection of Grant's activities at that time.
[3] See for example Ruse (1996: 99). See Harte & North (2004) for the full story of the foundation of the University of London, later

University College London on the University's chartering in 1836.

[4] Sixty of Grant's lectures were published in *The Lancet* (1833-1834), see the bibliography below. These lectures were considered the state of the art at the time by Grant's advocates and critics alike.

[5] Chuckerbutty won Grant's gold medal for Comparative Anatomy and, among other distinctions, went on to become professor at the Medical College in Calcutta (Sen Gupta, 1970). Chuckerbutty attended Grant's funeral, although he himself died just a few weeks later (Anonymous, 1874).

[6] For more on this see Sloan (1985) and Desmond (1989).

[7] We will see below how this practice led to some difficulties over intellectual property and plagiarism.

[8] The British Museum (Natural History) is now known as the Natural History Museum.

[9] There is no date for this letter given in Cornish (1904), however, it was written early in Flower's time at University College.

[10] Tragically this microscope has been lost. It was stolen from University College London and has never been recovered.

[11] See hand-written note in Grant's hand, UCL Special Collection: College Archives.

[12] See Desmond 1989 pp. 228-235 for discussion of these accusations and their repercussions.

[13] Many of these reports were excluded from the bibliography published in Desmond & Parker (2006).

[14] See Desmond (1989:28, n.42; 1997:11).

[15] This lecture series consisted of 60 lectures in total. In Desmond & Parker (2006) these lectures are listed as separate publications. This is due to their diverse subject matter and the fact that they were published weekly in separate issues of *The Lancet,* all save the last ten which appeared together at the end of volume 22. They are listed here without their individual page numbers.

References

Anonymous, 1825a Dr. Grant on the existence of the pancreas in some species of the cuttle-fish tribe. *Edinburgh Philosophical Journal* 13: 197-198.

Anonymous, 1825b On the existence of a pancreas in the *Doris argo*. *Edinburgh Philosophical Journal* 13: 198.

Anonymous 1850: Biographical sketch of Robert Edmond Grant, M.D., F. R. S. L. & E. &c. Professor of Comparative Anatomy and Zoology in University College, London. *The Lancet* 56: 686-695.

Anonymous, 1874-1875: Obituary notices of fellows of the Royal Society. *Proceedings of the Royal Society* 23: *vi-x*.

Barlow, N. 1958: *The Autobiography of Charles Darwin*. London: Collins. Pp 253.

Bowler, P. 1996: *Life's Splendid Drama: Evolutionary Biology and the Reconstruction of Life's Ancestry, 1860-1940*. Chicago & London: The University of Chicago Press. Pp 515.

Carpenter, W. B. 1835-1836: On the structure and functions of the organs of respiration. *West of England Journal of Science and Literature* 1: 217-228, 279-87.

Carpenter, W. B. 1858: *Vegetable Physiology and Systematic Botany*. London: Bohn's Scientific Library. Pp 606.

Carpenter, W. B. 1888: *Nature and Man: Essays Scientific and Philosophical*. London: Kegan Paul & Co. Pp 483.

Cornish, C. J. 1904: *Sir William Henry Flower: A Personal Memoir*. London: Macmillan and Co Limited. Pp 274.

Desmond, A. 1984: *Robert E. Grant (1793-1874): The Science and Social Predicament of a Pre-Darwinian Transmutationist, Including a Study of Richard Owen's Response to the Threat Posed by Grant's Transformism.* Unpublished PhD thesis, University of London.

Desmond, A. 1989: *The Politics of Evolution: Morphology, Medicine, and Reform in Radical London*. Chicago and London: The University of Chicago Press. Pp 503.

Desmond, A. 1997: *Huxley: Evolution's High Priest*. London: Michael Joseph. Pp 370.

Desmond, A. & Parker, S. E. Forthcoming (2006): The bibliography of Robert Edmond Grant (1793-1874): Illustrated with a previously unpublished photograph. *Archives of Natural History*.

Endersby, J. 2003: Escaping Darwin's shadow. *Journal of the History of Biology* 36: 385-403.

Fleming, J. 1828 *A history of British animals, exhibiting the descriptive characters and systematical arrangement of the genera and species of quadrupeds, birds, reptiles, fishes, Mollusca, and Radiata of the United Kingdom; including the indigenous, extirpated, and extinct kinds, together with periodical and occasional visitants*. Edinburgh: Bell & Bradfute. Pp 565.

Fletcher, K. 2004: Flower, Sir William Henry (1831–1899). *Oxford Dictionary of National Biography*, Oxford University Press, [http://www.oxforddnb.com/view/article/9766, accessed 5 Jan 2006].

Flower, W. H. 1898: *Essays on Museums, and Other Subjects Connected with Natural History*. London and New York:

Macmillan. Pp 394.

Gardiner, B. G., 1999 Grant and Darwin. *The Linnean* 15: 5-15.

Geoffroy Saint-Hilaire, E. 1818: *Philosophie anatomique*. Paris: Mequignon-Marvis.

Harte, N. & North, J. 2004: *The World of UCL: 1828-2004*. UCL Press. Pp 304.

Hodge, M. J. S., 1985 Darwin as a lifelong generation theorist, pp 207-243 in KOHN, D. (editor), *The Darwinian heritage*. Princeton: Princeton University Press. Pp 1,152.

Jesperson, P. H., 1949 Charles Darwin and Dr. Grant. *Lychnos* 1: 159-167.

Lester, J. 1995: *E. Ray Lankester and the Making of Modern British Biology*. (Ed) P. Bowler. British Society for the History of Science Monographs. Pp 220.

Poore, G. V. 1901: Robert Edmond Grant. *University College Gazette* 2 (34):190-191.

Roget, P. M. 1846: Dr. Roget's Bridgewater treatise. *The Lancet* 47: 482-483.

Ruse, M. 1996: *Monad to Man: The Concept of Progress in Evolutionary Biology*. Cambridge, Massachusetts: Harvard University Press. Pp 628.

Sen Gupta, P. C. 1970: Soorjo Coomar Goodeve Chuckerbutty: the first Indian contributor to modern medical science. *Medical History* 14: 183-191.

Schäfer, E. A. 1901: William Sharpey. *University College*

Gazette 2 (36):215.

Sloan, P. R., 1985 Darwin's invertebrate program, 1826-1836: preconditions for transformism, pp 71-120 in KOHN, D. (editor) *The Darwinian heritage*. Princeton: Princeton University Press. Pp 1,152.

Smith, R. 2004: Carpenter, William Benjamin (1813–1885). *Oxford Dictionary of National Biography*, Oxford University Press, [http://www.oxforddnb.com/view/article/4742, accessed 6 Jan 2006].

Acknowledgements

This booklet was made possible by grants from the Royal Society and by the University of London Central Research Fund. Thanks also to Michael Ruse, Lucyle T. Werkmeist Professor of Philosophy at Florida State University for reviewing the booklet and Joe Cain, Helen Chatterjee, Jack Ashby and Janet Knell for comments and feedback on earlier drafts of this booklet.

About the Author

Sarah Parker is Assistant Curator at the Grant Museum of Zoology and Comparative Anatomy, UCL and researches into the history of the Museum and its founder. This booklet was produced as part of an ongoing project at the Grant Museum to compile a complete research resource to Robert Edmond Grant.